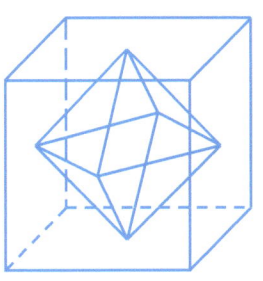
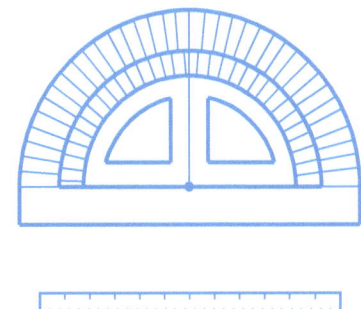
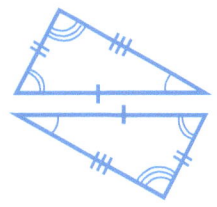

A Basic Course

in Geometry

– Part 5 of 5

Bill Lembke

Citrus Software Publishing

Citrus Ridge, Florida

Published by Citrus Software Publishing, a Division of Citrus Software Corporation

Printed and bound in the United States of America.

A Basic Course in Geometry – Part 5 of 5

ISBN-13: 978-1477574737

ISBN-10: 1477574735

Table of Contents – Part 1 of 5

Term	Definition
A Basic Course in Geometry	Geometry textbook (2013).
AAA	Angle-angle-angle; proof of spherical triangle congruency.
AAS	Angle-angle-side; proof of plain triangle congruency.
Acute Angle	Angle with a positive rotation of greater than 0 turns and less than 1/4 turn.
Acute Triangle	Triangle with all angles measuring less than 90 degrees.
Adjacent Angles	Angles that share a common vertex and edge but do not share any interior points.
Alternate Exterior Angles	Angles on opposite sides of a transversal line and outside of the parallel lines. These angles are equal in measure.
Alternate Interior Angles	Angles on opposite sides of a transversal line and inside of the parallel lines. These angles are equal in measure.
Altitude	A straight line through a vertex and perpendicular to the opposite side.
Angle	The space between two rays or line segments that meet at a common end point or vertex.
Angle Bisector	Transitive line that divides the angle into two equal parts.
Angle Side	One of the two rays forming an angle.
Annulus	Same as circular ring.
Annulus sector	Same as circular ring sector.
Antecedent	The 'if' part of a conditional; represented by p.
Antidipryramid	Same as a trapezohedron.
Antipodal points	A pairs of points on a sphere located on opposite ends of a diameter.
Antiprism	A prism with all lateral faces being triangles.
Apeirohedron	An infinite dimensional polytope or apeirotope.
Apeirotope	An infinite sided polytope or apeirohedron.
Apex	The highest point or vertex of a pyramid or cone; A pointed end.
Apothem	The distance from the center of a polygon to the midpoint of a side.
Approximately Equal (≈)	Numbers which are almost equal.
Arc	A curved portion of a circle.
Arc Length	The distance between an arc's endpoints along the path of the circle.
Archimedean Solids	Convex semiregular polyhedron with regular polygon faces, symmetrical edges, and symmetrical vertices. Unlike a Platonic solid, the faces are made up of two or more different polygons. Each polyhedron is isogonal and convex.
Archimedes	Greek mathematician.
Area	Measurement of the interior space of a polygon.

Argument	The collection of reasons or evidence used to prove a conjecture.
ASA	Angle-side-angle; proof of plain triangle congruency; proof of spherical triangle congruency.
Asymptote Lines	Lines that form the boundary of the hyperbola curve.
Augmented	Descriptive term or adjective that is part of a polyherdon name. A pyramid or cupola is joined to a face of a solid.
Augmented Dodecahedron	One of ninety-two Johnson solids.
Augmented Hexagonal Prism	One of ninety-two Johnson solids.
Augmented Pentagonal Prism	One of ninety-two Johnson solids.
Augmented Sphenocorona	One of ninety-two Johnson solids.
Augmented Triangular Prism	One of ninety-two Johnson solids.
Augmented Tridiminished Icosahedron	One of ninety-two Johnson solids.
Augmented Truncated Cube	One of ninety-two Johnson solids.
Augmented Truncated Dodecahedron	One of ninety-two Johnson solids.
Augmented Truncated Tetrahedron	One of ninety-two Johnson solids.
Auxiliary Lines	Lines that are added to a given drawing which help to demonstrate something and prove a statement of a proof.
Axiom	A self-evident assumption.
Ball	The area enclosed by a sphere.
Base	A designated side or sides from which to reference measurements of an object.
Base Angle	The angle opposite one of the equilateral sides in an isosceles triangle.
Betweenness of Points	A term that refers to the order of three collinear points. If A, B, and C are three different collinear points, point C is between points A and B if $AC + CB = AB$.
Bi- or Di-	Descriptive term or adjective that is part of a polyherdon name. Two copies of a solid are joined base to base.
Biangle	Two sided shape on a sphere.
Biaugmented Pentagonal Prism	One of ninety-two Johnson solids.
Biaugmented Triangular Prism	One of ninety-two Johnson solids.
Biaugmented Truncated Cube	One of ninety-two Johnson solids.
Bicentric	Polygon possessing both an incircle and a circumcircle.
Bicentric Quadrilateral	Quadrilateral that is both cyclic and tangential.
Bifrustum	Two frustra joined together.
Bigyrate Diminished Rhombicosidodecahedron	One of ninety-two Johnson solids.
Bilunabirotunda	One of ninety-two Johnson solids.
Bipyramid	A polyhedron created by attaching two pyramids symmetrically base to base. A regular square bipyramid is also called an octahedron.
Bisect	To divide into two equal parts.

Bisectrix	A line that bisects an angle.
Brahmagupta	Indian mathematician.
Brahmagupta's formula	A formula used to calculate the area of a cyclic quadrilateral.
Capsule	A three-dimensional stadium or a cylinder with two hemispherical caps.
Catalan Solids	Dual polyhedrons to the Archimedean solids. The Catalan solids are convex and isohedral. Unlike the Platonic solids and the Archimedean solids, the faces are not regular polygons. Each face is identical.
Catalan, Eugene	Belgian mathematician.
Cells	The bounding polyhedra of a polychoron.
Center of a Circle	The point that all points in the circle are equidistant from.
Center of Rotation	The point where the two intersecting lines of a rotation meet.
Central Angle	Angle at the center of a polygon made by two adjacent radius lines; Angle whose vertex is at the center of the circle; Angle formed at the center of the sphere by the intersection of two great circles; Also called a dihedral angle or spherical angle.
Centroid	The three medians intersect in a single point of a triangle.
Chiliagon	1000-gon.
Chord	Line segment whose end points lie on the circle.
Circle	One of four types of conics; The set of all points in a plane at an equal distance from a fixed point; A special case of an ellipse in which the major axis and minor axis have the equal lengths.
Circular Arc	The measurement of the rotation of one side of an angle from the standard position.
Circular Protractor	Protractor in the shape of a circle.
Circular Ring	A ring shaped object that is the region lying between two concentric coplanar circles.
Circular Ring Sector	Part of a ring shaped object that is the region lying between two concentric coplanar circles.
Circular Sector	Part of a circle bounded by two radii and their intercepted arc. It is a wedge shaped piece of a circle.
Circular Segment	Part of a circle bounded by a chord and its associated arc.
Circumcenter	Center of the circumcircle; The three perpendicular bisectors meet in a single point in a triangle; It equidistant from the three vertices and the common distance is the radius of the circumcircle.
Circumcircle	A circle outside of a polygon that intersects each of the vertices.
Circumference	Perimeter of a circle; The length of any great circle; The intersection of the sphere with any plane passing through its center.

Circumradius	Radius of a cirumcircle.
Circumscribed Sphere	The three-dimensional analogue of the circumscribed circle. A circumscribed sphere of a polyhedron is a sphere that contains the polyhedron and touches each of the polyhedron's vertices.
Circumscribed Triangle	Triangle that is inside of a circumcircle.
Closed Curve	The result when the last line segment is attached to the first line segment.
Collinear	In the same line.
Common Notation	Same as axiom.
Compass	One of four important geometry tools used in geometric construction; Used to construct circles or arcs.
Compass Only or Mohr-Mascheroni Construction	One of six types of geometric construction; Compass does not have markings, collapses after each use, and can not transfer distance.
Complementary Angles	Two angles that sum to one right angle.
Complex Polygon	Polygon with a self-intersecting boundary.
Complex Polyhedron	One of three types of polyhedron; The surface is not uniformly flat or is like a star.
Complex Quadrilateral	One of three types of quadrilaterals; It is self-intersecting. This shape is called a butterfly, bow tie, or hour glass.
Compound Polyhedron	A polyhedron composed of a number of interpenetrating polyhedra, either of the same of several different types, which share a common center.
Concave Polyhedron	One of three types of polyhedron; It has a hole or indentation.
Concave Quadrilateral	One of three types of quadrilaterals; It has two pairs of adjacent equal sides and is concave. This shape is called a dart or arrow.
Concave Simple Polygon	Simple polygon with at least one interior angle that exceeds 180 degrees.
Concentric Circles	Circles in the same plane that have the same center but have radii of different lengths.
Conditional	A statement that tells if one thing happens, another will follow; Written as if p, then q
Cone	Conic solid with a circular base; A surface of revolution generated by an oblique line that rotates around a fixed point, at a fixed angle from the axis, with both the line and axis passing through that fixed point.
Cone Frustum	Frustum created from a cone.
Congruent ($\cong$)	Geometric objects, such as angles and polygons, which have the same shape and size.
Congruent Arcs	Arcs in the same circle or congruent circles that have the same degree measure.
Congruent Marks	Short lines used to indicate congruent lines or angles.
Congruent Triangles	Triangles with all pairs of corresponding angles congruent

	and all pairs of corresponding sides have the same length. This is a total of six equalities. Congruent triangles have the exact same shape and size. If triangles are congruent, then they are also similar.
Conic Section	A curve resulting from the intersection of a right circular cone and a plane.
Conic Solid	A three-dimensional geometric shape bounded by a plane base and the surface formed by line segments connecting the perimeter of the base to a common point, or apex, outside the plane of the base.
Conical Frustum	A shape with properties common to both a cone and a cylinder. A conical frustum can be constructed by cutting off the top of a cone with a plane parallel to the base or by reducing the diameter of one base of a cylinder.
Conical Surface	A double cone that is generated by rotating a line in the Y-Z plane about the Z axis with the vertex at the origin.
Conjecture	An unproven proposition or hypothesis that is believed to be true.
Conjugate Axis	The y-axis in a hyperbola; Axis perpendicular to the transverse axis and passes through the center.
Connected Mathematics Project (CPM)	A problem-centered teaching approach based on the NCTM standards developed by Michigan State University.
Consecutive Interior Angles	Angles on the same side of a transversal line and inside of the parallel lines.
Consecutive Sides	Sides of a polygon that share an endpoint.
Consecutive Vertices	The endpoints of a single side of a polygon.
Consequent	The 'then' part of a conditional; represented by q.
Construction	A precise way of drawing which allows only two tools, the straightedge and the compass.
Contradiction	A statement that is always false. If x=1 is a true statement, then the statement x≠1 is a contradiction or false statement.
Convex Polyhedron	One of three types of polyhedron; Its surface does not intersect itself and a line segment drawn to connect any two points of the polyhedron is contained in the interior or on the surface.
Convex Quadrilateral	One of three types of quadrilaterals; All sides are convex.
Convex Simple Polygon	Simple polygon with no internal angles greater than 180 degrees.
Coordinate Axis	Two mutually perpendicular that divide a plane into four regions.
Coplanar	Within the same plane.
Corresponding Angles	Angles in matching position at an intersection of lines. These angles at the point of intersection are equal in size.
Cosia-	Polygon name prefix for 100.

Coterminal Angles	Two angles that have the same terminal side.
Counterexample	A situation in a conditional for which the antecedent is true, but the conditional is false; Same as contradiction.
CPCTC	Corresponding parts of congruent triangles are congruent.
Cross-section	A cross-section of a space figure is the shape of a particular two-dimensional "slice" of a space figure.
Cube	One of five Platonic solids; Made up of six squares; An equilateral square prism; Three-dimensional polytope; One of seven hexahedra; A regular hexahedron.
Cubic Prism	A prism with a square or rectangular base.
Cuboctahedron	One of thirteen Archimedean solids
Cupolae	Descriptive term or adjective that is part of a polyherdon name; Cup-shaped.
Curriculum and Evaluation Standards	Publication by the NCTM (1989).
Cyclic Convex Polygon	Convex polygon with all of the vertices lying on a single circle.
Cyclic Quadrilateral	A quadrilateral that can be circumscribed.
Cylinder	A three-dimensional non-polytope constructed from two congruent circular bases located in parallel planes and the lateral surface connecting the two bases.
Decacross	10 dimensional polytope or Polyxennon.
Decagon	10-gon.
Decahedron	Polyhedron with 10 faces or sides.
Decayotton	9 dimensional polytope or Polyyotton.
Decimal Notation	One of two methods for expressing degrees, listing degrees and converting minutes and seconds to a combined decimal value of the fractional degree.
Definition	A statement of precise meaning.
Degenerate Conic	A conic created when a plane passes through the vertex or intersects one nappe parallel to the side of the cone, resulting in a point or line.
Degree (°)	Angle measurement with 360 parts, or degrees, making a complete circle of rotation.
Degree Protractor	Protractor that measures in degrees.
Degree-Minute-Second Notation	One of two methods for expressing degrees, listing degrees, minutes, and seconds.
Dekeract	10 dimensional polytope or Polyxennon.
Deltoidal Hexecontahedron	One of thirteen Catalan solids.
Deltoidal Icositetrahedron	One of thirteen Catalan solids.
Depth	The volume distance of an object on an intersecting perpendicular plane; Z-axis.
Detachment	One of two forms of deductive reasoning; Uses one conditional statement and a statement to reach a conclusion.

Di-	Polygon name prefix for 2.
Diagonal	A line connecting two vertices that are not a side; A segment in a polygon whose endpoints are two nonconsecutive vertices.
Diameter	A chord that passes through the center of the circle; The longest chord of a circle; A chord that passes through the center of the sphere.
Dihedral Angle	Angle at which two faces meet.
Dimension	The minimum number of coordinates needed to specify each point within it.
Diminished	Descriptive term or adjective that is part of a polyherdon name; A pyramid or cupola is removed from a solid.
Diminished Rhombicosidodecahedron	One of ninety-two Johnson solids.
Dipyramid	Same as bipyramid.
Direct Proof	A type of proof in which the conclusion is shown to be true directly from statements of the proof. Direct proofs are the most common type of proof.
Directrix	A line outside of an ellipse that is parallel to the minor axis.
Disdyakis Dodecahedron	One of thirteen Catalan solids.
Disdyakis Triacontahedron	One of thirteen Catalan solids.
Disk	The area enclosed by a circle.
Disphenocingulum	One of ninety-two Johnson solids.
Dodecagon	12-gon.
Dodecahedron	One of five Platonic solids; Made up of twelve pentagons; Polyhedron with 12 faces or sides.
Dot	A visual representation of a point.
Double Cone	Two cones placed apex to apex.
Doubly Infinite Cone	The union of any set of straight lines that pass through a common apex point, and therefore extends symmetrically on both sides of the apex. This kind of cone does not have a bounding base and extends to infinity.
Dual Polyhedron	Every polyhedron is uniquely related to another specific polyhedron. The polyhedron and the dual polyhedron have the same number of edges, but the vertices and faces are reversed or occupy complementary locations.
Edge	The bounding line segments of a polyhedron.
Elements	Book written by Euclid. It is an arrangement of 465 propositions covering not only plane and solid geometry but also much of what is now known as algebra, trigonometry, and advanced arithmetic. One of the most important geometry books.
Ellipse	One of four types of conics; The set of all points in a plane in which the sum of the distances from two fixed points is constant. This set of points forms a continuous closed

	curved line.
Elongated	Descriptive term or adjective that is part of a polyherdon name; A prism is joined to the base of a solid or between the bases of a solid.
Elongated Pentagonal Cupola	One of ninety-two Johnson solids.
Elongated Pentagonal Dipyramid	One of ninety-two Johnson solids.
Elongated Pentagonal Gyrobicupola	One of ninety-two Johnson solids.
Elongated Pentagonal Gyrobirotunda	One of ninety-two Johnson solids.
Elongated Pentagonal Gyrocupolarotunda	One of ninety-two Johnson solids.
Elongated Pentagonal Orthobicupola	One of ninety-two Johnson solids.
Elongated Pentagonal Orthobirotunda	One of ninety-two Johnson solids.
Elongated Pentagonal Orthocupolarotunda	One of ninety-two Johnson solids.
Elongated Pentagonal Pyramid	One of ninety-two Johnson solids.
Elongated Pentagonal Rotunda	One of ninety-two Johnson solids.
Elongated Square Cupola	One of ninety-two Johnson solids.
Elongated Square Dipyramid	One of ninety-two Johnson solids.
Elongated Square Gyrobicupola	One of ninety-two Johnson solids.
Elongated Square Pyramid	One of ninety-two Johnson solids.
Elongated Triangular Cupola	One of ninety-two Johnson solids.
Elongated Triangular Dipyramid	One of ninety-two Johnson solids.
Elongated Triangular Gyrobicupola	One of ninety-two Johnson solids.
Elongated Triangular Orthobicupola	One of ninety-two Johnson solids.
Elongated Triangular Pyramid	One of ninety-two Johnson solids.
Endpoints	An endpoint is a point used to define a line segment or ray. A line segment has two endpoints, a ray has one.
Ennea-	Polygon name prefix for 9.
Enneaconta-	Polygon name prefix for 90.
Enneacontagon	90-gon.
Enneacross	9 dimensional polytope or Polyyotton.
Enneadecagon	19-gon.
Enneagon	9-gon.
Enneazetton	8 dimensional polytope or Polyzetton.
Enneract	9 dimensional polytope or Polyyotton.
Equal (=)	Numbers that have the same value.
Equiangular	All angles are congruent.
Equiangular triangle	Triangle with three angles of equal measurement.
Equidistant	To be equal distance from two points.
Equilateral	All sides of equal length.
Euclid	Greek mathematician.

Euler, Leonhard	Swiss mathematician.
Euler's polyhedron formula	Formula that defines the number of faces, edges, and vertices of a spherical polyhedron.
Expansion	One of four processes used to modify Platonic solids to form Archimedean solids; The faces are moved outward from the center of a polyhedron, with the vertices connected across the empty spaces.
Explementary Angles	Two angles that sum to a full angle.
Exterior Angle	An angle formed on the outside of a polygon between a side and the extended adjacent side
Exterior Angles	Angles formed between the two lines that are crossed by the transversal line and outside of the parallel lines.
Exterior of a Circle	The set of all points whose distance from the center is greater than the length of the radius of the circle.
Exterior of a Sphere	The set of all points whose distance from the center is greater than the length of the radius of the sphere.
Face	The bounding polygons of a polyhedron.
Fermat Point	A point of a triangle that is a minimum distance from the three vertices.
Finite	With limit or countable.
Focus	Each of the two fixed points of an ellipse.
Frustum	A polyhedron created by cutting off the top of a conic solid with a plane parallel to the base of the solid. The portion between the top plane and the base plane is the frustum.
Frustum Height	The perpendicular distance between the planes of the two bases.
Full Angle	Angle with a positive rotation of 1 turn.
General Triangle	A triangle that represents all triangles. This means that descriptions, properties, and formulas that apply to a general triangle apply to all triangles, regardless of size or angles.
Generator	The oblique line used to generate a cone.
Geodesic	The arc of the great circle between the two points; The shortest distance, a curved line, between any two points on a sphere, is an arc of the great circle through the two points.
Geometric Center	In a regular triangle, center of the inscribed and circumscribed circles.
Geometric Construction	The drawing of lengths, angles, and objects on a plane using drawing tools and rules that govern construction techniques.
Geometric Proof	A series of statements based on definitions, postulates, axioms, and theorems used to show that a mathematical statement is valid or true.
Geometry	A branch of mathematics which studies spatial

	relationships and spatial structures. It is concerned with the properties and relationships of points, lines, angles, curves, surfaces, and solids.
Given	Information assumed to be true in a proof.
Great Circle	A circle created by the intersection of a plane and a sphere, with the plane passing through the center of the sphere.
Great Dodecahedron	One of four Kepler-Poinsot solids.
Great Icosahedron	One of four Kepler-Poinsot solids.
Great Stellated Dodecahedron	One of four Kepler-Poinsot solids.
Gyrate	Descriptive term or adjective that is part of a polyherdon name; A cupola on a solid is rotated so different edges match up.
Gyrate Bidiminished Rhombicosidodecahedron	One of ninety-two Johnson solids.
Gyrate Rhombicosidodecahedron	One of ninety-two Johnson solids.
Gyro-	Descriptive term or adjective that is part of a polyherdon name; Joined with unlike faces.
Gyrobifastigium	One of ninety-two Johnson solids.
Gyroelongated	Descriptive term or adjective that is part of a polyherdon name; An antiprism is joined to the base of a solid or between the bases of a solid.
Gyroelongated Pentagonal Bicupola	One of ninety-two Johnson solids.
Gyroelongated Pentagonal Birotunda	One of ninety-two Johnson solids.
Gyroelongated Pentagonal Cupola	One of ninety-two Johnson solids.
Gyroelongated Pentagonal Cupolarotunda	One of ninety-two Johnson solids.
Gyroelongated Pentagonal Pyramid	One of ninety-two Johnson solids.
Gyroelongated Pentagonal Rotunda	One of ninety-two Johnson solids.
Gyroelongated Square Bicupola	One of ninety-two Johnson solids.
Gyroelongated Square Cupola	One of ninety-two Johnson solids.
Gyroelongated Square Dipyramid	One of ninety-two Johnson solids.
Gyroelongated Square Pyramid	One of ninety-two Johnson solids.
Gyroelongated Triangular Bicupola	One of ninety-two Johnson solids.
Gyroelongated Triangular Cupola	One of ninety-two Johnson solids.
HA	Hypotenuse-angle; Proof of plain triangle congruency, for a right triangle.
Half-plane	The set of points in a plane that lie on one side of a line.
Hebesphenomegacorona	One of ninety-two Johnson solids.
Hectagon	100-gon.
Height	The vertical distance on a plane; Y-axis.
Hemicube	One of seven hexahedra; A cube with a plane cutting through two opposite corners and the midpoint of two edges.

Hemiobelisk	One of seven hexahedra; An elongated square pyramid or obelisk with one of the four base corners of the square pyramid cut off to create a new triangular face.
Hemiprism	Descriptive term or adjective that is part of a polyherdon name; A half prism where the cutting plane contains only a single vertex, instead of a whole edge.
Hemisphere	Two equal segments of a sphere created by a great circle; Half of a sphere
Hena-	Polygon name prefix for 1.
Hendecagon	11-gon.
Hendecaxennon	10 dimensional polytope or Polyxennon.
Hepta-	Polygon name prefix for 7.
Heptaconta-	Polygon name prefix for 70.
Heptacontagon	70-gon.
Heptacross	7 dimensional polytope or Polyexon.
Heptadecagon	17-gon.
Heptagon	7-gon.
Heptahedron	Polyhedron with 7 faces or sides.
Heptapeton	6 dimensional polytope or Polypeton.
Hepteract	7 dimensional polytope or Polyexon.
Heron	Greek mathematician.
Hexa-	Polygon name prefix for 6.
Hexaconta-	Polygon name prefix for 60.
Hexacross	6 dimensional polytope or Polypeton.
Hexadecachoron	4 dimensional polytope or polychoron.
Hexadecagon	16-gon.
Hexagon	6-gon.
Hexagonal Antiprism	Antiprism with a hexagon base.
Hexahedron	Polyhedron with 6 faces or sides.
Hexateron	5 dimensional polytope or Polyteron.
Hexecontrahedron	Polyhedron with 60 faces or sides.
Hexeract	6 dimensional polytope or Polypeton.
Hextacontagon	60-gon.
Hidden Lines	Broken lines used to signify lines that normally wouldn't be seen in a drawing.
Hilbert, David	German mathematician.
HL	Hypotenuse-leg; Proof of plain triangle congruency, for a right triangle.
Hollow Shape Polyhedron	One of three definitions of polyhedron based on bounding characteristics; A boundary between the internal and external areas formed by the faces of polygons. The volume of the internal space is not included. It is a hollow shape.
Hyperbola	One of four types of conics; The set of all points in a plane in that the difference between each of two fixed points and

	any point on the hyperbola is constant.
Hypercube	One of three special classes of regular convex polytope that exist in every dimensionality; A generalization of square to an arbitrary dimension. The sides are in parallel pairs orthogonal or perpendicular to each other.
Hyperplane	A generalization of the concept of a plane for multidimensional objects, such as points, lines, and planes.
Hypotenuse	The side opposite the right angle in a triangle.
Icosahedron	One of five Platonic solids; Up of twenty equilateral triangles; Polyhedron with 20 faces or sides.
Icosidodecahedron	One of thirteen Archimedean solids; Polyhedron with 32 faces or sides.
Icositetrahedron	Polyhedron with 24 faces or sides.
Idealized Compass	A compass that creates perfect circles.
Idealized Straightedge	A straightedge that creates perfectly straight and of infinite length.
Incenter	Center of the incircle; The three angle bisectors intersect in a single point in a triangle; It is equidistant from the three sides and the common distance is the radius of the incircle.
Incircle	A circle inside of a polygon that touches each of the sides.
Included Angle	The angle between the two lines of the angle; The angle made by two sides of a polygon.
Included Side	The side between two angles in a polygon.
Indirect Proof	A type of proof in which the conclusion is shown to be true indirectly from statements of the proof. The statements show that all of the alternatives to the conclusion are shown to be false. The conclusion is shown to be true because the assumption that its negation is true leads to a contradiction.
Infinite	Without limit or uncountable.
Initial Side	The side of the angle on the x-axis.
Inradius	Radius of an incircle; The same as the apothem.
Inscribed Triangle	Triangle that contains an incircle.
Inscriptible Quadrilateral	A quadrilateral that can be inscribed.
Interior Angle	Angle formed by two adjacent sides inside the polygon.
Interior Angles	Angles formed between the two lines that are crossed by the transversal line and inside of the parallel lines.
Interior of a Circle	The set of all points whose distance from the center is less than the length of the radius of the circle.
Interior of a Sphere	The set of all points whose distance from the center is less than the length of the radius of the sphere.
Intersecting Planes	Planes that share a line.
Intersection	The term intersect is used when lines, rays, line segments or figures meet, that is, they share a common point. The

	point they share is called the point of intersection.
Inverse	A form of conditional; If not p, then not q.
Irrational Number	Decimal number that never ends and never repeats, such as pi (π).
Irregular Polygon	Polygon that does not have equal side lengths and congruent angles.
Irregular Prism	A prism with an irregular polygon base.
Irregular Pyramid	A pyramid with an irregular polygon base.
Isogon	Polygon that is equilateral and equiangular; Also called a regular polygon.
Isogonal	One of three types of polyhedron symmetry; A vertex-transitive polyhedron has symmetrical vertices. This means that each vertex is surrounded by the same kinds of face in the same or reverse order and with the same angles between the corresponding faces.
Isohedral	One of three types of polyhedron symmetry; A face-transitive polyhedron has symmetrical faces. This means that all of the faces must be congruent.
Isometry	Transformation of an object into another object by rotations (turning), reflections (flipping), or translations (sliding).
Isosagon	20-gon.
Isosceles Trapezoid	Quadrilateral with 1 pair of parallel sides and 1 pair of equal sides; One of eight convex quadrilaterals.
Isosceles Triangle	A triangle with at least two side of equal length.
Isosi-	Polygon name prefix for 20.
Isotoxal	One of three types of polyhedron symmetry; A edge-transitive polyhedron has symmetrical edges. This means that there is only one type of edge to an object, such as a hexagon face meeting another hexagon face. The dihedral angle is the same for all edges.
Johnson Solids	Convex polyhedra with regular polygon faces and equal edge lengths, but are not isogonal. Unlike a Platonic solid, the faces are made up of two or more different polygons.
Johnson, Norman	American mathematician.
Kai-	Polygon name prefix for "and".
Kepler, Johannes	German mathematician.
Kepler-Poinsot Solids	Regular polyhedron with regular polygon faces, symmetrical edges, and symmetrical vertices. Each polyhedron is isogonal, isotoxal, isohedral, and concave (stellated). The only four regular polyhedra which are not convex.
Kite	Quadrilateral with 2 pairs of adjacent equal sides; One of eight convex quadrilaterals.
LA	Leg-angle; Proof of plain triangle congruency, for a right triangle.

Lateral Edges	The segments where the lateral faces meet in a pyramid.
Lateral Faces	Triangular shaped of sides of a pyramid.
Lateral Surface	The side of a cone.
Leg	A side of a right triangle that include the 90 degree angle.
Lembke, Bill	American mathematician.
Line	A series of points extending infinitely in a straight path in two opposite directions.
Line Segment	Part of a line, with points marking the beginning and ending locations; 1 dimensional polytope or polytelon.
Linear Pair	Two supplementary adjacent angles whose non-common sides form a line.
LL	Leg-leg; Proof of plain triangle congruency, for a right triangle.
Locus	The set of all points, and only those points, that satisfy a given condition.
Magnitude of Rotation	The amount of rotation in degrees.
Major Arc	An arc that lies in the exterior of the central angle that intercepts the arc.
Major Axis	The longest diameter of an ellipse.
Mascheroni, Lorenzo	Italian mathematician.
Mathematical Pun	Without geometry, life is pointless.
Meridian	Any great circle passing through a point designated a pole.
Meta-	Descriptive term or adjective that is part of a polyherdon name; The solid has two oblique faces augmented. Oblique faces are not right angles or a multiple of right angle.
Metabiaugmented Dodecahedron	One of ninety-two Johnson solids.
Metabiaugmented Hexagonal Prism	One of ninety-two Johnson solids.
Metabiaugmented Truncated Dodecahedron	One of ninety-two Johnson solids.
Metabidiminished Icosahedron	One of ninety-two Johnson solids.
Metabidiminished Rhombicosidodecahedron	One of ninety-two Johnson solids.
Metabigyrate Rhombicosidodecahedron	One of ninety-two Johnson solids.
Metagyrate Diminished Rhombicosidodecahedron	One of ninety-two Johnson solids.
Minor Arc	An arc that lies in the interior of the central angle that intercepts the arc.
Minor Axis	Shortest diameter of an ellipse.
Minute or Arc Minute (′)	Angle measurement with 60 parts, or minutes, making a degree.
Modern Construction	One of six types of geometric construction; All drawing tools have markings. Compass can be fixed, can transfer distance.

Modern Traditional Construction	One of six types of geometric construction; Straightedge and compass have markings. Compass can be fixed, can transfer distance.
Modified Traditional Construction	One of six types of geometric construction; Straightedge and compass do not have markings. Compass can be fixed, can transfer distance.
Mohr, George	Danish mathematician.
Monad	0 dimensional polytope, such as point.
Morley Center	The intersection point of the line segments connecting the Morley triangle vertices with the opposite original triangle vertices.
Morley Circle	The circumcircle of the Morley triangle.
Morley Triangle	An equilateral triangle formed by the three points of intersection of the adjacent angle trisectors.
Myriagon	10000-gon.
Nappe	Each of the two cones of a double cone.
National Council of Teachers of Mathematics (NCTM)	Organization that establishes mathematic standards for pre-K-12.
n-cube	An n-dimensional hypercube, where n can represent any number 0 or larger.
Negative Angle	Clockwise, or downward, rotation from the terminal side.
n-gon	Generalized name for a polygon, where n can represent any number 0 or larger; a polygon with n sides.
Noble Polyhedron	One of six categories of polyhedron based on symmetrical properties; Isogonal and Isohedral.
Nonahedron	Polyhedron with 9 faces or sides.
Non-cyclic Convex Polygon	A convex polygon with all of the vertices not lying on a single circle.
Non-included Side	The side of a triangle that is not included by two given angles.
Non-polytope	Objects created from curved lines.
Non-star Shape Complex Polygon	A complex polygon not in the shape of a star.
Non-uniform Polyhedron	One of six categories of polyhedron based on symmetrical properties; Not isogonal, every face is regular polygon.
n-orthoplex	An n-dimensional orthoplex, where n can represent any number 0 or larger.
n-simplex	An n-dimensional simplex, where n can represent any number 0 or larger.
n-tope	Generalized name for a polytope, where n can represent any number 0 or larger.
Oblique Angle	An angle that is not a right angle, or a multiple of a right angle.
Oblique Conical Frustum	A conical frustum where the central axis does not intersects the bases perpendicularly.
Oblique Cylinder	A cylinder where the central axis does not intersects the

	bases perpendicularly.
Oblique Frustum	A frustum where an axis through the center of the upper base and the lower base does not intersects the bottom base perpendicularly.
Oblique Lines	Intersecting lines that are not perpendicular to one another.
Oblique Prism	A prism where the lateral faces are not perpendicular to the bases. The lateral faces are parallelograms, but not rectangles.
Oblique Pyramid	A pyramid where an axis through the vertex and the center of the base, the altitude, does not intersect the base perpendicularly.
Oblique Wedge	A wedge where non-parallel triangular sides slant symmetrically towards the center.
Obtuse Angle	Angle with a positive rotation greater than 1/4 turn and less than 1/2 turn.
Obtuse Triangle	A triangle with one angle measuring greater than 90 degrees.
Octa-	Polygon name prefix for 8.
Octaconta-	Polygon name prefix for 80.
Octacontagon	80-gon.
Octacross	8 dimensional polytope or Polyzetton.
Octadecagon	18-gon.
Octaexon	7 dimensional polytope or Polyexon.
Octagon	8-gon.
Octahedron	One of five Platonic solids; Made up of eight equilateral triangles; Polyhedron with 8 faces or sides; A regular square bipyramid; Three- dimensional polytope.
Octeract	8 dimensional polytope or Polyzetton.
One Dimension	Length only.
One Letter	One of three naming methods of an angle. Using a single letter that corresponds to the vertex, provided that this does not cause any confusion.
One Number	One of three naming methods of an angle. Placing a number at the vertex and in the interior of the angle.
Opposite Faces	Faces that lie in parallel planes.
Opposite Rays	Two rays with a common endpoint that form a line.
Orientation	In an image change, the direction in which the points named move (i.e., how A's position relates to B's and B's relates to C's); Either clockwise or counterclockwise for figures.
Origin Point	The intersection of the x-axis, y-axis, and z-axis at the center of the rectangular coordinate system.
Ortho-	Descriptive term or adjective that is part of a polyherdon name; Joined with like faces.

Orthocenter	three altitudes intersect in a single point in a triangle.
Orthodiagonal	The diagonals form perpendicular or right angles in a quadrilateral.
Orthodome	The shortest path between two non-antipodal points on the surface of a sphere.
Orthogonal Lines	Lines that meet at a right angle; Also called perpendicular lines
Orthoplex	One of three special classes of regular convex polytope that exist in every dimensionality; A generalization of square to an arbitrary dimension. The vertices are in pairs orthogonal or perpendicular to each other.
Orthoprism	A prism with all lateral faces being rectangles or squares.
Outside Angle	Angle formed by two adjacent sides outside of the polygon.
Oval	A shape consisting of a closed curve that may or may not be symmetrical. It is unlike a circle or ellipse and does not have a precise mathematical definition.
Overlapping Triangles	Triangles that share a side or angles.
Para-	Descriptive term or adjective that is part of a polyherdon name; The augmented faces are parallel.
Parabiaugmented Dodecahedron	One of ninety-two Johnson solids.
Parabiaugmented Hexagonal Prism	One of ninety-two Johnson solids.
Parabiaugmented Truncated Dodecahedron	One of ninety-two Johnson solids.
Parabidiminished Rhombicosidodecahedron	One of ninety-two Johnson solids.
Parabigyrate Rhombicosidodecahedron	One of ninety-two Johnson solids.
Parabola	One of four types of conics; The set of all points in a plane in that are equidistance from a fixed line and a fixed point not on the line.
Paragraph Proof	A type of proof format in which the steps are written out in complete sentences and arranged in a paragraph. It is less formal than a two-column proof, but identical in content.
Paragyrate Diminished Rhombicosidodecahedron	One of ninety-two Johnson solids.
Parallel Lines	Lines that are always the same distance apart and do not intersect; Two or more coplanar lines that have no points in common.
Parallel Planes	Planes that have no points in common.
Parallelogram	Quadrilateral with 2 pairs of parallel sides; One of eight convex quadrilaterals.
Pasch, Moritz	German mathematician.
Penta-	Polygon name prefix for 5.
Pentachoron	4 dimensional polytope or polychoron.

Pentaconta-	Polygon name prefix for 50.
Pentacontagon	50-gon.
Pentacross	5 dimensional polytope or Polyteron.
Pentadecagon	15-gon.
Pentagon	Two-dimensional polytope; 5-gon.
Pentagonal Cupola	One of ninety-two Johnson solids.
Pentagonal Dipyramid	One of ninety-two Johnson solids.
Pentagonal Gyrobicupola	One of ninety-two Johnson solids.
Pentagonal Gyrocupolarotunda	One of ninety-two Johnson solids.
Pentagonal Hexecontahedron	One of thirteen Catalan solids.
Pentagonal Icositetrahedron	One of thirteen Catalan solids.
Pentagonal Orthobicupola	One of ninety-two Johnson solids.
Pentagonal Orthobirotunda	One of ninety-two Johnson solids.
Pentagonal Orthocupolarotunda	One of ninety-two Johnson solids.
Pentagonal Pyramid	One of ninety-two Johnson solids; One of seven hexahedra; A pyramid whose base is a pentagon.
Pentagonal Rotunda	One of ninety-two Johnson solids.
Pentagonal Wedge	One of seven hexahedra; A tetrahedron with two corners cut off.
Pentahedron	Polyhedron with 5 faces or sides.
Pentakis Dodecahedron	One of thirteen Catalan solids.
Penteract	5 dimensional polytope or Polyteron.
Percentage Protractor	Protractor that measures in percentages.
Perimeter	The measurement around the outside of an object.
Perpendicular Bisector	A bisector where a straight line passing through the midpoint of a side and being perpendicular to it, forms a right angle with it.
Perpendicular Lines	Lines that meet at a right angle; Also called orthogonal lines.
Perpendicular Planes	Planes in which any two intersecting lines, one in each plane, form a right angle.
Pi (π)	The value of a circle's' circumference divided by the diameter; Equal to 3.141592654 to nine decimal places.
Plane	A flat surface extending infinitely in all directions.
Plane Figures	Figures that lie on a plane.
Plane Geometry	The study of figures on a flat surface.
Plato	Greek mathematician.
Platonic Solids	Polyhedra with regular polygon faces, symmetrical edges, and symmetrical vertices. Each polyhedron is isogonal, isotoxal, isohedral, and convex; The only five convex regular polyhedra.
Poinsot, Louis	French mathematician.
Point	A location; 0 dimensional polytope or monad.
Polychoron	4 dimensional polytope, such as Pentachoron,

	Hexadecachoron, Tesseract.
Polyexon	7 dimensional polytope, such as Octaexon, Heptacross, Hepteract.
Polygon	A two-dimensional object bounded by straight lines; A two-dimensional polytope or a simple closed polygonal curve on a plane; A closed plane figure bounded by three or more line segments.
Polygon Boundary	The line segments that separate the interior of a polygon from the exterior of a polygon. It defines the polygon shape.
Polygon Exterior	The outside of a polygon.
Polygon Interior	The inside of a polygon.
Polygonal Chain	In polytope creation, line segments attached like a links in a chain, end to end.
Polygonal Curve	In polytope creation, when two line segments meet at each vertex; Also called a topological curve.
Polyhedral Angle	The angle of a polyhedron between the segments joining the center and the vertices.
Polyhedral Surface	The faces of a polyhedron.
Polyhedron	A three-dimensional shape bounded by faces (polygons), edges (line segments), and vertices (points). Since a polyhedron is comprised of polygons, it has flat faces and straight edges; A three-dimensional object bounded by polygons; 3 dimensional polytope.
Polyhedron Circumradius	The radius of sphere circumscribed around a polyhedron.
Polyhedron Net	A flat pattern that can then be folded along the edges and taped together to regenerate the polyhedron; A two-dimensional figure that can be folded on its segments or curved on its boundaries to form a three-dimensional figure.
Polyhedron Symmetry	A concept of balance and self-similarity. A polyhedron can be rotated around its center point along the X, Y, and Z axis. After a rotation, a symmetrical polyhedron should appear similar to the original view.
Polypeton	6 dimensional polytope, such as Heptapeton, Hexacross, Hexeract.
Polytelon	1 dimensional polytope, such as Line Segment.
Polyteron	5 dimensional polytope, such as Hexateron, Pentacross, Penteract.
Polytope	An object created from straight lines; A finite region of n-dimensional space, where n is an arbitrary number, enclosed by a finite number of hyperplanes. It is a multidimensional solid with flat sides.
Polyxennon	10 dimensional polytope, such as Hendecaxennon, Decacross, Dekeract.
Polyyotton	9 dimensional polytope, such as Decayotton, Enneacross,

	Enneract.
Polyzetton	8 dimensional polytope, such as Enneazetton, Octacross, Octeract.
Poncelet, Jean Victor	French mathematician.
Positive Angle	Counterclockwise, or upward, rotation from terminal side.
Postulate	A statement that is accepted as true without proof.
Principles and Standards for School Mathematics	Publication by the NCTM (2000).
Prism	A polyhedron with two identical *n*-sided polygonal bases and *n* other parallelogram lateral faces connecting the corresponding sides of the bases. The two bases have the same size and shape.
Proof	A sequence of logical deductions based on accepted assumptions and previously proven statements that verify statement is true.
Protractor	One of four important geometry tools used in geometric construction; A flat curved ruler with markings that correspond to distances, usually measured in degrees or radians, for measuring angles.
Pyramid	A polyhedron in the form of a conic solid; A conic solid with a polygon base is a pyramid. All pyramids are self-dual.
Pyramid Frustum	A frustum created from a pyramid.
Pyramid Height	The perpendicular distance from the vertex to the base of a pyramid.
Pyramid Vertex	The apex where all of the sides meet in a pyramid.
Pythagoras	Greek mathematician.
Pythagorean Theorem	A formula for determining the lengths of the sides of a right triangle, in relation to each other.
Q.E.D.	An abbreviation that stands for the Latin *quod erat demonstrandum*, meaning "that which was to have been demonstrated." It is translated from the Greek όπερ έδει δειξαι, meaning "precisely what was required to be proved."
Q.E.F.	An abbreviation that stands for the Latin *quod erat faciendum*, meaning "that which was to have been done." It is translated from the Greek όπερ έδει ποιησαι, meaning "precisely what was required to be done."
Qin Jiushao	Chinese mathematician.
Quadrangle	4-gon; Same as quadrilateral or tetragon.
Quadrants	Four regions of a plane created by the coordinate axis.
Quadrilateral	A polygon with four sides and four angles; 4-gon; same as tetragon or quadrangle.
Quasi-convex Polyhedron	A type of polyhedron derived by tunneling into a convex polyhedron to remove a section.
Quasi-regular Polyhedron	One of six categories of polyhedron based on symmetrical

	properties; Isogonal and, Isotoxal; Every face is regular polygon.
Radian	The measure of an angle created by an arc that has the same length as the circle's radius. The angle in radians can also be thought of as the ratio between the arc length and the radius. A complete circle rotation is 2π radians.
Radian protractor	Protractor that measures in radians.
Radius	The distance from the center of the polygon to any vertex; The distance from the center of the circle to a point on the circle; The distance from the center of the sphere to a point on the sphere.
Ray	Half of a line. It has a point marking the beginning location and extends infinitely in one direction.
Rectangle	Quadrilateral with 2 pairs of parallel sides and all angles equal to 90 degree angles; One of eight convex quadrilaterals
Rectangular Coordinate System	A measurement system in which two mutually perpendicular lines, called coordinate axis, intersect at a point called the origin which divide the plane into four regions called quadrants.
Rectangular Solid	A solid with six sides, all of which are rectangles.
Rectification	One of four processes used to modify Platonic solids to form Archimedean solids; A form of truncation with one point cut on an edge at each vertex, such that the vertex is cut off.
Reflection	A type of isometry; The flipping of an object.
Reflex Angle	Angle with a positive rotation of greater than 1/2 turn and less than 1 turn.
Regular Polygon	Polygon that is equilateral and equiangular; Also called an isogon.
Regular Polyhedron	One of six categories of polyhedron based on symmetrical properties; Isogonal, Isotoxal, and Isohedral; Every face is same regular polygon.
Regular Prism	A prism with a regular polygon base.
Regular Pyramid	A pyramid with a regular polygon base.
Regular Triangle	A polygon with three equal sides and three congruent angles.
Rhombic Dodecahedron	One of thirteen Catalan solids.
Rhombic Triacontahedron	One of thirteen Catalan solids.
Rhombicosidodecahedron	One of thirteen Archimedean solids.
Rhombicuboctahedron	One of thirteen Archimedean solids.
Rhombus	Quadrilateral with 2 pairs of parallel sides and 4 equal sides; One of eight convex quadrilaterals.
Right Angle	Angle with a positive rotation of 1/4 turn.
Right Antiprism	A prism where each of the triangles is an isosceles triangle.

Right Cone	A cone with a circular base that is perpendicular to its altitude.
Right Conical Frustum	A conical frustum where the central axis intersects the bases perpendicularly.
Right Cylinder	A cylinder where the central axis intersects the bases perpendicularly; A cylinder consisting of two circular bases that are perpendicular to its altitude.
Right Frustum	A frustum where an axis through the center of the upper base and the lower base intersects the bottom base perpendicularly.
Right Prism	A prism which has one base aligned directly above the other base. The lateral faces are perpendicular to the bases and rectangular.
Right Pyramid	A pyramid where an axis through the vertex and the center of the base, the altitude, intersects the base perpendicularly.
Right Rectangular Wedge	A wedge with parallel triangular sides.
Rotation	A type of isometry; The turning of an object.
Rotundae	Descriptive term or adjective that is part of a polyherdon name; Round-shaped.
Ruler	One of four important geometry tools used in geometric construction; A straightedge with distance markings.
SAS	Side-angle-side; Proof of plain triangle congruency; Proof of spherical triangle congruency.
Scalene Quadrilaterals	Quadrilaterals that do not have any special properties, and generally do not have parallel sides, equal sides, or congruent angles; One of eight convex quadrilaterals.
Scalene Triangle	A triangle with all three sides of unequal length.
Secant Line	A line that intersects a circle in two different points.
Second or Arc Second (″)	Angle measurement with 60 parts, or seconds, making a minute and 360 parts making a degree.
Semicircle	Half of a circle
Semi-circular Protractor	Protractor in the shape of a semi-circle.
Semi-disk	Half of a disk.
Semi-perimeter	Half of the distance of the perimeter.
Semi-regular Polyhedron	One of six categories of polyhedron based on symmetrical properties; Isogonal; Every face is regular polygon.
Sexagesimal System	System of numerals using the number 60 as its base; used in a modified form for measuring time, angles, and the geographic coordinates.
Side	A line segment that forms the edge of the polygon; The bounding line segments of a polygon.
Side Length	The length of the side of a polygon.
Similar (~)	Objects which have the same shape but not the same size.
Similar Triangles	Triangles with corresponding angles congruent or the corresponding sides have lengths that are in the same

	proportion; Similar triangles have the exact same shape, but may not be the same size.
Simple Curve	A curve that does not self-intersect.
Simple Polygon	Polygon with a single, non-intersecting boundary.
Simplex	One of three special classes of regular convex polytope that exist in every dimensionality; A generalization of a triangle to an arbitrary dimension. An n-simplex is an n-dimensional polytope with n + 1 vertices.
Skew Lines	Non-coplanar lines that do not intersect.
Slant Height	Distance from the vertex to the base, measured along a lateral face of a pyramid; the length of a lateral edge of a conic solid.
Small Circle	The circle formed by the intersection of a sphere and a plane that does not contain the center.
Small Stellated Dodecahedron	One of four Kepler-Poinsot solids.
Snub Cuboctahedron	One of thirteen Archimedean solids.
Snub Disphenoid	One of ninety-two Johnson solids.
Snub Dodecahedron	One of thirteen Archimedean solids.
Snub Square Antiprism	One of ninety-two Johnson solids.
Snubification	One of four processes used to modify Platonic solids to form Archimedean solids; A form of expansion of a polyhedron in which all of the faces are slightly rotated in the same direction. The empty spaces are filled with triangles.
Solid Geometry	The study of figures in three dimensions.
Solid Shape Polyhedron	One of three definitions of polyhedron based on bounding characteristics; the polygon surfaces and the internal volume. When describing a polyhedron in general, this definition is almost always used, as opposed to the wire frame and hollow shape definitions.
Space Figure	A space figure or three-dimensional figure is a figure that has depth in addition to width and height.
Sphenocorona	One of ninety-two Johnson solids.
Sphenoid	Descriptive term or adjective that is part of a polyherdon name; A wedge or half prism that includes the base polygon and its featured edge.
Sphenomegacorona	One of ninety-two Johnson solids.
Sphere	The set of all points in three-dimensional space at an equal distance from a fixed point. This set of points forms a continuous closed surface. A perfectly round geometrical object in three-dimensional space.
Spherical Angle	The central angle of a sphere.
Spherical Biangle	An object created by the intersection of two great circles on a sphere.
Spherical Cap	The region of a sphere which lies above, or is cut off by, a plane.

Spherical Excess	The amount by which the sum of the angles of a polygon on a sphere exceeds the sum of the angles of a polygon with the same number of sides in a plane.
Spherical Frustum	Same as spherical segment.
Spherical Geometry	The study of two-dimensional figures on the surface of a sphere.
Spherical Lune	The surface of a spherical wedge.
Spherical Polygon	A closed two-dimensional object on the surface of a sphere created by the intersection of two or more great circles.
Spherical Quadrilateral	An object created by the intersection of four great circles.
Spherical Sector	The region of a sphere bounded by two radii and their intercepted angle. It is a cone shaped piece of a sphere.
Spherical Segment	The region of a sphere which lies between two parallel planes.
Spherical Triangle	An object created by the intersection of three great circles.
Spherical Wedge	A portion of a sphere bounded by two plane semi-disks and their intercepted angle.
Spherical Zone	The surface of a spherical segment, excluding the two bases.
Square	Quadrilateral with 2 pairs of parallel sides, all angles equal to 90 degrees, and 4 equal sides; one of eight convex quadrilaterals; the only regular quadrilateral; 2 dimensional polytope.
Square Cupola	One of ninety-two Johnson solids.
Square Gyrobicupola	One of ninety-two Johnson solids.
Square Orthobicupola	One of ninety-two Johnson solids.
Square Pyramid	One of ninety-two Johnson solids.
Square Root	A number r whose square (the result of multiplying the number by itself, or $r \times r$) is x. For example, 4 is a square root of 16 because $4^2 = 16$.
SSA	Side-side-angle; Proof of plain triangle congruency
SSS	Side-side-side; Proof of plain triangle congruency; Proof of spherical triangle congruency.
Stadium	A shape consisting of a rectangle with semi-circles attached to two opposite ends.
Standard Position	An angle with its vertex at the origin point and initial side along the positive x-axis.
Star Pyramid	A pyramid with a star-shaped polygon base.
Star Shape Complex Polygon	Complex polygon in the shape of a star.
Steiner, Jacob	Swiss mathematician.
Stellation	The process of extending edges or faces until they meet to form a new polygon or polyhedron.
Stereometry	The measurement of volumes and dimensions of various solid figures including cylinder, circular cone, truncated

	cone, sphere, and prisms.
Stewart Toroidal Polyhedron	A toroidal polyhedron or solid with regular polygon faces.
Straight Angle	Angle with a positive rotation of 1/2 turn.
Straightedge	One of four important geometry tools used in geometric construction; used to construct straight lines and does not have any distance markings.
Straightedge Only or Poncelet-Steiner construction	One of six types of geometric construction; Straightedge does not have markings.
Supplementary Angles	Two angles that sum to a straight angle.
Surface	The boundary of a three-dimensional figure.
Surface Area	The total area of the surface of a solid.
Syllogism	One of two forms of deductive reasoning; Uses two conditional statements and reaches a conclusion by combining the hypothesis of one statement with the conclusion of another statement.
Symmetric Figure	A figure that can be folded flat along a line so that the two halves match perfectly.
Symmetry Orbit	The region of the circumscribed sphere.
Tangent	The touching of objects or lines; intersecting at one point.
Tangent Line	A line that intersects a circle in exactly one point.
Tangential Quadrilateral	A quadrilateral is inscribed so the circle is tangent or touching each side.
Terminal Side	The side of the angle that rotates away from the x-axis.
Tessellation	The process using shapes to fill an area, like tiling, with no overlapping shapes or gaps.
Tesseract	4 dimensional polytope or polychoron.
Tetra-	Polygon name prefix for 4.
Tetraconta-	Polygon name prefix for 40.
Tetracontagon	40-gon.
Tetradecagon	14-gon.
Tetradecahedron	Polyhedron with 14 faces or sides.
Tetragon	4-gon; Same as quadrangle or quadrilateral.
Tetragonal Antiwedge	One of seven hexahedra; A skewed pentagonal pyramid and is the least symmetric of the hexahedra. Its dual is a shape of its mirror own image.
Tetrahedron	One of five Platonic solids; Made up of four equilateral triangles; Polyhedron with 4 faces or sides; Three-dimensional polytope.
Tetrakis Hexahedron	One of thirteen Catalan solids.
The Foundation of Geometry	Book written by David Hilbert.
The Geometer's Sketchpad ®	Geometry software.
Theorem	A conjecture proven to be true.
Therefore, etc.	An abbreviation that stand for the exact restatement of the original proposition as the concluding statement of the proof.

Three-dimensional non-polytopes	Geometric objects with three dimensions that do not have straight sides. The sides of non-polytopes are curved.
Three dimensions	Length, width, and height.
Three Letters	One of three naming methods of an angle. The center letter corresponds to the vertex of the angle and the other letters representing points on the sides of the angle.
Topological Curve	In polytope creation, when two line segments meet at each vertex; also called a polygonal curve.
Toroid	The solid contained by the surface of a torus.
Toroidal Polyhedron	A polyhedron with one or more holes. All of the sides are composed of polygons.
Toroidal Space	The name used to describe the surface area and volume of a torus.
Torus	A surface formed by revolving a circle in three-dimensional space around the z-axis. The surface is ring-shaped with a hole, similar to a three-dimensional circular ring.
Traditional or Euclidian Construction	One of six types of geometric construction; Straightedge and compass do not have markings. Compass collapses after each use, can not transfer distance.
Transitive Line	A line that divides an object into two equal parts.
Translation	A type of isometry; the sliding of an object.
Transversal Line	A line that crosses two or more lines in the same plane.
Transverse Axis	The x-axis in a hyperbola; perpendicular to the conjugate axis and passes through the center.
Trapezohedron	A polyhedron with faces composed of congruent kites. A trapazohedron is similar to a bipyramid, but instead of triangular sides, it has kite sides. However, instead of attaching the two halves symmetrically like in a bipyramid, the two halves of a trapezohedron are slightly off set to allow the vertices to match up.
Trapezoid	Quadrilateral with 1 pair of parallel sides; one of eight convex quadrilaterals.
Tri-	Polygon name prefix for 3.
Triacontagon	30-gon.
Triakis Icosahedron	One of thirteen Catalan solids.
Triakis Octahedron	One of thirteen Catalan solids.
Triakis Tetrahedron	One of thirteen Catalan solids.
Triangle	Two dimensional polytope; The only rigid polygon; 3-gon; Same as trilateral or trigon.
Triangle 30-60-90	One of two special or common right triangles, based on degree of angles of 30, 60, and 90.
Triangle 45-45-90	One of two special or common right triangles, based on degree of angles of 45, 45, and 90.
Triangle Altitude	The triangle height. The altitude is perpendicular from the base to the opposite vertex. The altitude line divides a

	triangle into two opposite facing right triangles.
Triangle Base	Any one of the three sides of a triangle, but is usually drawn at the bottom.
Triangle Median	A line in a triangle from a vertex to the midpoint of the opposite side that divides the base into two equal parts.
Triangle Rigidity	In a triangle, if the lengths of the sides are known, then the angles can not be changed by shifting the sides and the length of less than all of the sides can not be changed without changing the angles.
Triangular Cupola	One of ninety-two Johnson solids.
Triangular Dipyramid	One of ninety-two Johnson solids; One of seven hexahedra; The dual of a triangular prism, and looks like two tetrahedral glued on a common face.
Triangular Hebesphenorotunda	One of ninety-two Johnson solids.
Triangular Orthobicupola	One of ninety-two Johnson solids.
Triangular Prism	A prism with a triangular base.
Triaugmented Dodecahedron	One of ninety-two Johnson solids.
Triaugmented Hexagonal Prism	One of ninety-two Johnson solids.
Triaugmented Triangular Prism	One of ninety-two Johnson solids.
Triaugmented Truncated Dodecahedron	One of ninety-two Johnson solids.
Triconta-	Polygon name prefix for 30.
Tricontahedron	Polyhedron with 30 faces or sides.
Tridiminished Icosahedron	One of ninety-two Johnson solids.
Tridiminished Rhombicosidodecahedron	One of ninety-two Johnson solids.
Trigon	3-gon; Same as triangle or trilateral.
Trigonometry	A subset of geometric principles concerned with the measurement of triangles.
Trigyrate Rhombicosidodecahedron	One of ninety-two Johnson solids.
Trilateral	3-gon; same as trigon or triangle.
Trisectors	Two lines that divide an angle into three equal angles.
Triskaidecagon	13-gon.
Truncated Cube	One of thirteen Archimedean solids
Truncated Cuboctahedron	One of thirteen Archimedean solids
Truncated Dodecahedron	One of thirteen Archimedean solids
Truncated Icosahedron	One of thirteen Archimedean solids
Truncated Icosidodecahedron	One of thirteen Archimedean solids
Truncated Octahedron	One of thirteen Archimedean solids
Truncated Tetrahedron	One of thirteen Archimedean solids
Truncation	One of four processes used to modify Platonic solids to form Archimedean solids; Two points are cut on an edge at each vertex, such that the vertex is cut off.
Turn	A rotation of the side of angle around the vertex.
Two dimensional non-polytopes	Geometric objects with two dimensions that do not have

	straight sides. The sides of non-polytopes are curved.
Two dimensions	Length and width.
Two-Column Proof	A type of proof format in which statements are listed in one column and the corresponding reasons for each statement being of true are listed in a second column. It is more formal than a paragraph proof, but identical in content.
Undecahedron	Polyhedron with 11 faces or sides.
Uniform Polyhedron	One of six categories of polyhedron based on symmetrical properties; Regular, Quasi-regular, or Semi-regular.
Unmeasured Angle	Angle without measurement used for illustrative purposes.
Van Hiele, Pierre and Dina	Dutch educators that created model for geometry instruction and student learning.
Vertex	Common endpoint of the rays forming an angle; The intersection of two sides at a point.
Vertical Angles	Angles that are formed when two lines intersect and form four angles. They are diagonal from each other at the intersection. These angles are equal in measure.
Vertices	The bounding end points line segments of a polygon or polyhedron.
Vertices of a Spherical Polygon	The points of intersection of the great circles. The sides of a spherical polygon are the geodesic segments between the points.
Volume	Measurement of the interior space of a polyhedron.
Wantzel, Pierre	French mathematician.
Wedge	A polyhedron created by cutting a prism with a plane containing some edge of either base polygon, but not intersecting the other base, resulting in a half prism.
Width	The horizontal distance on a plane; X-axis.
Wire Frame Polyhedron	One of three definitions of polyhedron based on bounding characteristics; A polyhedron as only the edges of the polygons. There is no surface area or internal volume. It resembles a wire frame.
X Axis Rotation	Rotation along the x-axis around the center of a polyhedron.
X-axis	Axis with horizontal measurement; Also east and west, or left and right.
Y Axis Rotation	Rotation along the y-axis around the center of a polyhedron.
Y-axis	Axis with vertical measurement; Also north and south, or forward and back.
Z Axis Rotation	Rotation along the z-axis around the center of a polyhedron.
Z-axis	Axis with measurement perpendicular to both the x-axis and y-axis; Also up and down.
Zero Angle	Zero angle rotation from the standard position; the initial

	side and the terminal sides of the angle coincide.
Zero Dimension	Without length, width, or height.

Appendix B – Chapter Test Solutions

Chapter 1 – Concepts and Standards

Section 1 – Geometry Concepts

1.2 – The Geometry Universe

1. C   2. B   3. A   4. B   5. A   6. C   7. C   8. A   9. B

1.3 – Definitions

1. I   2. E   3. C   4. L   5. A   6. G   7. K   8. H   9. D   10. F   11. J   12. B

1.4 – Geometry Objects

1. A, C, Circle   2. B, D, Cube   3. B, C, Square   4. A, D, Sphere

1.5 – Geometry Measurements

1. A = 12   B = 12   C= 14   D = 12   Object: C   Rule: Perimeter = 12.
2. E = 16   F = 16   G = 16   H = 18   Object: H   Rule: Perimeter = 16.
3. I = 20   J = 18   K = 16   L = 12   Object: J   Rule: Perimeter is divisible by 4.
4. M = 10   N = 24   O = 18   P = 12   Object: M   Rule: Perimeter is divisible by 6.

Section 2 – Geometry Standards

1.8 – Van Hiele Levels of Geometric Understanding

1. B   2. D   3. A   4. E   5. C

1.9 – Progress Indicators

1. C   2. A   3. D   4. B

1.10 – Learning Phases

1. A   2. E   3. C   4. B   5. D

1.11 – Connected Mathematics Project

1. B   2. C   3. A   4. C   5. A   6. B

Chapter 2 – Angles

2.2 – Coordinate System

Coordinates of points (X, Y)

Part 1 – Identify Quadrant

1. IV	2. IV	3. III	4. Y	5. Y	6. Y	7. X
8. II	9. X	10. X,Y	11. I	12. X	13. I	14. II
15. II	16. II	17. III	18. III	19. III	20. IV	21. IV
22. I	23. X	24. I	25. Y			

Part 2 – Write Coordinates

1. (5, -2)	2. (8,-8)	3. (-3, -3)	4. (0,8)	5. (0, -6)	6. (0, 4)
7. (0, 5)	8. (-4,2)	9. (-3, 0)	10. (0, 0)	11. (2, 2)	12. (9, 0)
13. (2, 6)	14. (-2, 4)	15. (-3, 6)	16. (-6, 3)	17. (-9, -4)	18. (-6, -6)
19. (-4, -9)	20. (2, -5)	21. (5, -5)	22. (6, 2)	23. (-7, 0)	24. (4, 4)
25. (0,-2)					

Part 3 – Identify Points

1. A 2. E 3. J 4. O 5. T 6. Y 7. Q 8. G 9. W 10. M 11. C 12. N 13. I

14. X 15. D 16. S 17. U 18. K 19. B 20. F 21. V 22. L 23. R 24. P 25. H

Part 4 – Mark Points

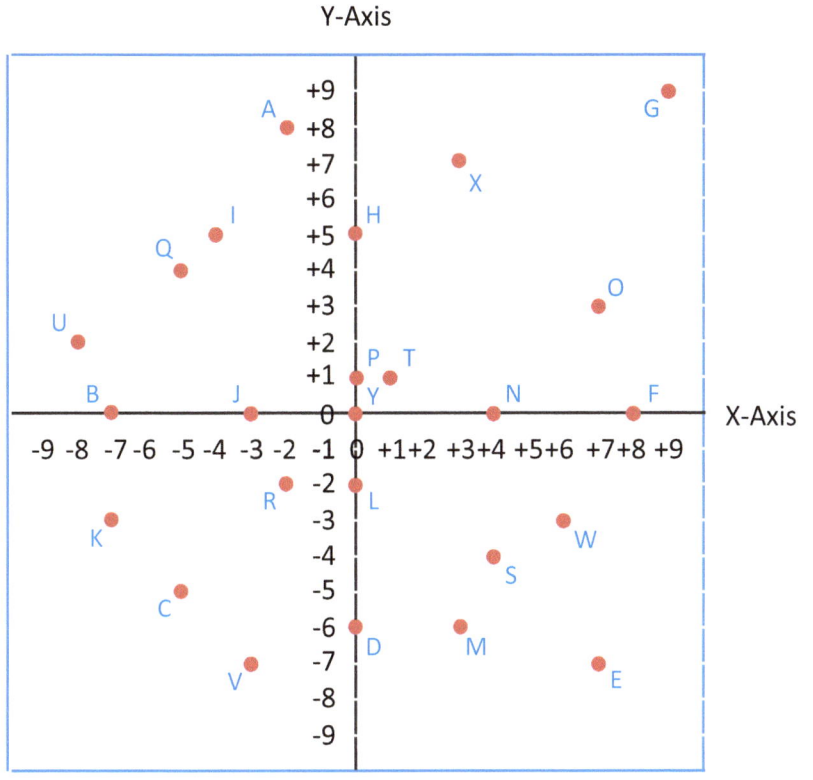

Mark and label each of the 25 points.				
1. A (-2, 8)	2. B (-7, 0)	3. C (-5, -5)	4. D (0, -6)	5. E (7, -7)
6. F (8, 0)	7. G (9, 9)	8. H (0, 5)	9. I (-4, 5)	10. J (-3, 0)
11. K (-7, -3)	12. L (0, -2)	13. M (3, -6)	14. N (4, 0)	15. O (7, 3)
16. P (0, 1)	17. Q (-5, 4)	18. R (-2, -2)	19. S (4, -4)	20. T (1, 1)
21. U (-8, 2)	22. V (-3, -7)	23. W (6, -3)	24. X (3, 7)	25. Y (0, 0)

2.3 – Labeling Angles

1. EAH, HAE, 1, A   2. FCG, GCF, 2, C   3. EBF, EBI, IBF   4. HDG, HDK, KDG

5. EIF, EIB, BIF   6. EJG, EJH, HJF, HJG, GJF, FJE

7. AHD, AHG, AHJ, AHE, EHD, EHG, EHJ, JHD, JHG, GHD

2.4 – Equivalence and Congruence

Translations – Draw the triangle after translation. (Original triangle in blue, translation in red.)

Part 1 – Rotation

1. Rotate the triangle $+90^0$.

2. Rotate the triangle -180^0.

3. Rotate the triangle $+270^0$.

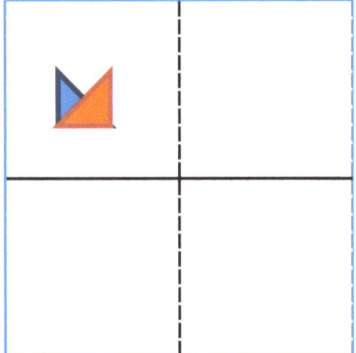 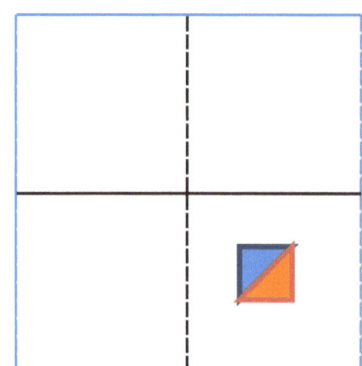 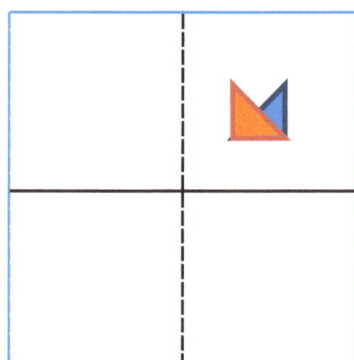

Part 2 – Reflection

4. Reflect the triangle over the x-axis.

5. Reflect the triangle over the y-axis.

6. Reflect the triangle over the x-axis.

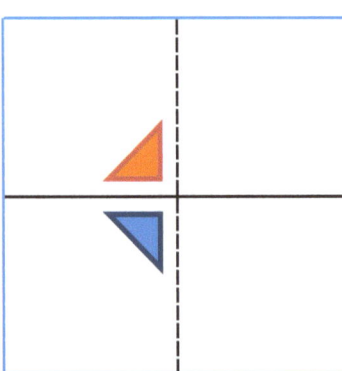 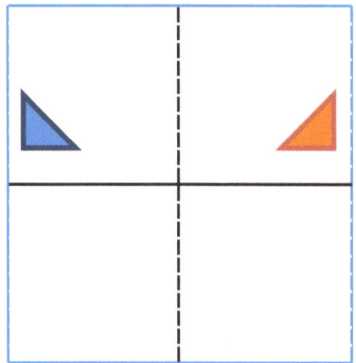 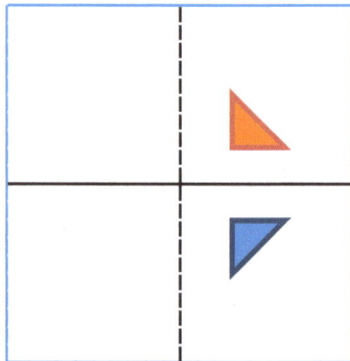

Part 3 – Translation

7. Translate the triangle left 5 units, then up 2 units.

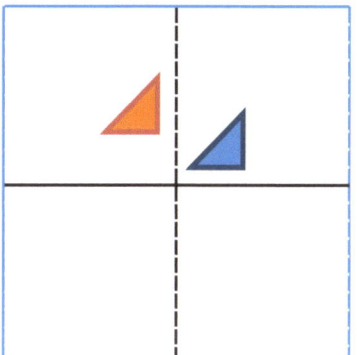

8. Translate the triangle right 3 units, then down 4 units.

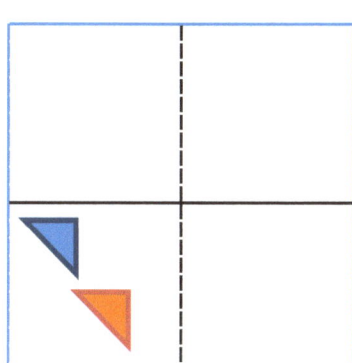

9. Translate the triangle right 6 units, then up 6 units.

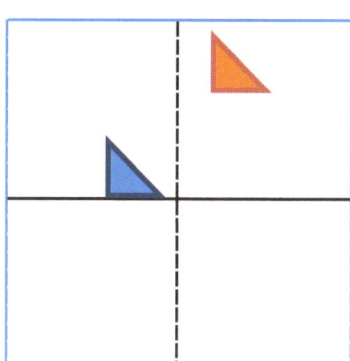

Part 4 – Combination

10. Rotate the triangle -90⁰, then reflect over the x-axis.

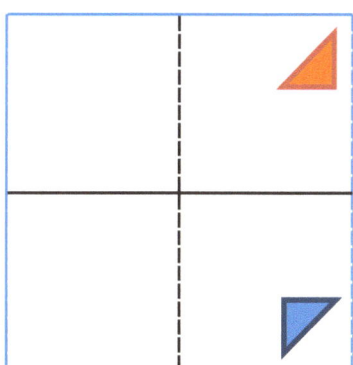

11. Reflect over the y-axis, then translate 3 left, 2 up.

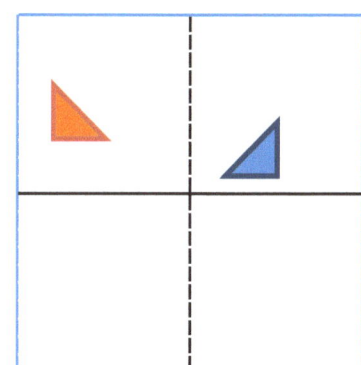

12. Translate right 2, up 4, then rotate -270⁰.

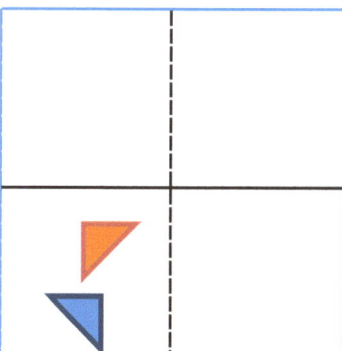

2.5 – Angle Measurement

1. 90   2. 15   3. 120 4. 30   5. 105 6. 75   7. 45   8. 60

2.6 – Degrees

1. 25.75^0 2. 82.505^0 3. 120.4^0 4. 180.2125^0 5. 275.1^0

6. 317^0 52' 30"   7. 9^0 7' 30"   8. 46^0 21' 36"

9. 75^0 40' 12"   10. 153^0 28' 48"

2.7 – Radians

1. 0.3491   2. 1.3090   3. 2.7925   4. 3.6652   5. 5.5851

6. 42.9718   7. 143.2394   8. 186.2113   9. 257.8310   10. 329.4507

2.8 – Angle Types

1. right   2. full   3. zero   4. reflex   5. acute

6. full   7. straight   8. obtuse   9. straight   10. right

2.9 – Angle Comparison

1. 30   2. 45   3. 75   4. 120   5. EXA, AXF   6. GXD, BXH, HXC

7. EXB, FXB   8. BXC, CXA, AXD

Chapter 3 – Polytopes

3.1 – Introduction

1. C   2. E   3. A   4. D   5. B   6. G   7. H   8. J   9. I   10. F

3.2 – Types of Polytopes

1. A   2. E   3. D   4. C   5. B

3.3 – Regular Convex Polytopes

1. B   2. A   3. C   4. A   5. C   6. B

3.4 – Polytopes – Simplex, Orthoplex, and Hypercube

1. E   2. C   3. J   4. A   5. I   6. F   7. B   8. H   9. G   10. D

3.5 – Polytope Decomposition

1. C   2. F   3. A   4. G   5. D   6. B   7. E

Chapter 4 – Polygons

4.2 – Polygon Classification
1. F   2. D   3. A   4. C   5. E   6. B

4.3 – Polygon Parts
1. 120, 300   2. 90, 270   3. 72, 252   4. 60, 240

4.4 – Polygon Formulas

1. 1260, 27, 7   2. 1440, 35, 8   3. 1620, 44, 9   4. 1800, 54, 10

5. 140, 220, 40   6. 144, 216, 36   7. 147.27, 212.73, 32.72   8. 150, 210, 30

4.5 – Polygon Names
1. tricontakaihexagon   2. tetracontakaioctagon   3. heptacontakaidigon

4. enneacontakaihenagon

4.6 – Regular Polygon Properties
1. 50   2. 15   3. 10.5   4. 7.5

5. 72   6. 60   7. 51.42   8. 45

4.7 – Triangle

1. I   2. F   3. B   4. D   5. G   6. A   7. H   8. C   9. E

4.8 – Triangle Formulas

1. 17.41, 1.65, 4.52   2. 27.71, 2.30, 4.61   3. 24, 2, 5

4.10   – Regular Polygon Areas

1. 403.0615   2. 9.0000   3. 84.2908   4. 27.5276   5. 98.9092

6. 64.9519   7. 120.7107   8. 14.5356   9. 15.5885

4.11   – Drawing Regular Polygons

1. 120   2. 90   3. 72   4.60   5. 51.428571

6. 45   7. 40   8. 36   9. 32.7272   10. 30

Chapter 5 – Triangles and Quadrilaterals

Section 1 – Triangles

5.2 – Triangle Types

1. 10.8253   2. 1.4434   3. 2.8868   4. 4.3301   5. 6.5450   6. 26.1799

7. 4.4721   8. 8.6603   9. 6.9282   10. 2.8284   11. 8.0000   12. 11.1803

13. 3.4641, 4.0000   14. 6.9282, 8.0000   15. 10.3923, 12.0000   16. 13.8564, 16.0000

17. 3.0000, 4.2426   18. 6.0000, 8.4853   19. 9.0000, 12.7279   20. 12.0000, 16.9706

5.3 – Similar and Congruent Triangles

1. similar, S1   2. congruent, LA   3. congruent, AAS   4. congruent, ASA

5. similar, S3   6. congruent, HL   7. congruent, SAS   8. congruent, LL

9. congruent, SSA   10. similar, S2   11. congruent, HA   12. congruent, SSS

5.4 – Points, Lines, and Circles of Triangles

1. E   2. B   3. D   4. A   5. F   6. C

5.5 – Triangle Rigidity

1. T   2. T   3. F   4. F

Section 2 – Quadrilaterals

5.8 – Convex Quadrilaterals

1. H   2. G   3. A, C, F, G   4. A, E, G   5. A, D, F

6. B, C 7. F, G   8. A, E, G

5.9 – Quadrilateral Formulas

1. 24   2. 20   3. 15   4. 25   5. 18   6. 30   7. 32   8. 20

Chapter 6 – Polyhedron

6.1 – Introduction

1. T   2. F   3. F   4. T   5. T

6.2 – Polyhedron Name Conventions

1. E   2. B   3. D   4. F   5. A   6. C

6.3 – Polyhedron Name – Number of Faces

1. D   2. B   3. G   4. E   5. H   6. A   7. C   8. F

6.4 – Polyhedron Name – Shape of Faces

1. H   2. B   3. D   4. F   5. E   6. C   7. G   8. A

6.5 – Polyhedron Symmetry

1. C   2. D   3. B   4. E   5. A

6.6 – Polyhedron Characteristics

1. B   2. E   3. C   4. A   5. D

6.7 – Euler's Formula

1. 12   2. 26   3. 150   4. 8   5. 24   6. 150

6.8 – Polyhedron Nets

1. F   2. F   3. T   4. T

Chapter 7 – Polyhedron Solids – Part 1

7.2 – Platonic Solids

1. 3.6742, 1.2247, 2.1213   2. 1.4433, 2.5000, 1.7678   3. 2.8284,1.6330, 2.0000

4. 4.2038, 3.3405, 3.9271   5. 1.9021, 1.5115, 1.6180

6. 10.3923, 25.4558   7. 150.0000, 125.0000   8. 13.8564, 4.6188

9. 45.1244, 206.9042   10.34.6410, 17.4536

7.3 – Kepler-Poinsot Solids

1. T   2. F   3. F   4. T

7.4 – Archimedean Solids

1. E   2. D   3. A   4. B   5. C

7.5 – Catalan Solids

1. T   2. F   3. T   4. T

7.6 – Johnson Solids

1. K   2. F   3. H   4. A   5. C   6. M   7. E   8. J   9. B   10. L   11. G   12. D   13. I

Chapter 8 – Polyhedron Solids – Part 2

8.2 – Pyramids

1. 36, 8   2. 81, 27   3. 144, 64   4. 225, 125   5. 324, 216   6. 576, 512

8.3 – Bipyramids

1. 40, 16   2. 90, 54   3. 160, 128   4. 250, 250   5. 360, 432   6. 640, 1024

8.4 – Trapezohedra

1. T   2. F   3. F   4. T

8.5 – Frusta

1. 60, 17.5   2. 120, 140   3. 270, 472.5   4. 468, 1120   5. 750, 2187.5   6. 1080, 3765

8.6 – Prisms

1. 160, 128   2. 360, 288   3. 640, 1024   4. 1000, 2000

5. 184, 120   6. 414, 270   7. 736, 960   8. 1150, 1875

8.7 – Antiprisms

1. T   2. F   3. F   4. F

8.8 – Wedges

1. 22.5   2. 75.9375   3. 180   4. 351.5625

5. 21   6. 70.875   7. 168   8. 328.125

8.9 – Toroidal Polyhedra

1. F   2. T   3. T   4. T

8.10 – Compound Polyhedra

1. T   2. F   3. T

Chapter 9 – Two Dimensional Non-polytopes

9.2 – Conic Sections

1. E   2. B   3. D   4. A   5. F   6. C

9.3 – Types of Conics

1. E   2. B   3. D   4. C   5. A

9.4 – Circles

1. H   2. F   3. K   4. M   5. C   6. G   7. A   8. P

9. E   10. O   11. I   12. D   13. N   14. J   15. B   16. L

17. 25.1327, 50.2655, 4.1888   18. 31.4159, 78.5398, 3.9270

19. 18.8496, 28.2743, 4.7124   20. 37.6991, 113.0973, 12.5664

21. 20.5664, 25.1327   22. 25.7080, 39.2699

23. 15.4248, 14.1372   24. 30.8496, 56.5487

9.5 – Circular Sectors

1. 12.1888, 8.3776   2. 24.5664, 37.6991   3. 13.9270, 9.8175   4. 10.7124, 7.0686

9.6 – Circular Segments

1. 8.1888, 1.4556   2. 22.8164, 8.0638   3. 7.7395, 0.4696   4. 8.9624, 1.0816

9.7 – Circular Rings

1. 28.2743   2. 84.8230   3. 47.1234   4. 113.0973

9.8 – Circular Ring Sectors

1. 14.1371   2. 7.0686   3. 9.8175   4. 4.7124

9.9 – Ellipses

1. 37.6991, 21.9911   2. 31.4159, 21.9911   3. 87.9646, 34.5575   4. 125.6637, 40.8407

9.10 – Parabolas

1. 12.0000   2. 13.3333   3. 6.6667   4. 16.0000

9.11 – Hyperbolas

1. F   2. D   3. B   4. E   5. C   6. A

9.12 – Stadiums

1. 11.1416, 14.2832   2. 44.5664, 28.5664   3. 64.2743, 30.8496   4. 130.2655, 45.1327

9.13 – Ovals

1. T   2. T   3. T   4. F

Chapter 10- Three Dimensional Non-polytopes

10.2 – Spheres

1. C   2. F   3. O   4. J   5. H   6. A   7. N   8. L

9. M   10. B   11. G   12. I   13. D   14. K   15. E

16. 113.0973, 113.0973   17. 201.0619, 268.0826

18. 314.1593, 523.5988   19. 452.3893, 904.7787

10.3 – Spherical Caps

1. 12.5664, 12.6318   2. 122.5221, 554.3275   3. 15.7080,11.4813   4. 14.1372, 21.5003

10.4 – Spherical Sectors

1. 100.5309, 16.7552   2. 490.0885, 245.0442   3. 107.0105, 26.1799   4. 93.3053, 14.1372

10.5 – Spherical Segments

1. 150.7964, 151.3200   2. 276.4602, 384.0597   3. 109.9557, 94.9023   4. 47.9093, 32.9867

10.6 – Spherical Wedges

1. 33.5103, 44.6804   2. 150.7964, 301.5929   3. 39.2699, 65.4498   4. 28.2743, 28.2743

10.7 – Cones

1. 21.9911, 4.1888   2. 51.8363, 14.1372   3. 87.9646, 33.5103   4. 141.3717, 65.4498

10.8 – Cylinders

1. 113.0973, 87.9646   2. 226.1946, 251.3274   3. 339.2920, 339.2920   4. 452.3893, 201.0619

10.9 – Conical Frusta

1. 100.5310, 81.6814   2. 135.0885, 238.7610   3. 87.1792, 44.7677   4. 355.0000, 270.1770

10.10 – Tori

1. 118.4353, 59.2176   2. 315.8273, 315.8273   3. 592.1763, 888.2644   4. 947.4820, 1894.9640

10.11 – Capsules

1. 37.6991, 16.7552   2. 150.7964, 134.0413   3. 226.1947, 282.7433   4. 452.3893, 770.7374

Chapter 11 – Spherical Geometry

11.2 – Terms

1. T, X	2. P, Y	3. G, Y	4. C, Z	5. N, X	6. R, X
7. A, Z	8. I, Y	9. D, X	10. B, Y	11. L, X	12. J, Y

11.3 – Properties

1. G   2. B   3. E   4. J   5. H   6. C   7. A   8. F   9. I   10. K   11. D

11.4 – Spherical Polygons

1. E   2. I   3. A   4. C   5. F   6. D   7. H   8. B   9. G

11.5 – Spherical Angles

1. 36, 15.7080, 314.1593   2. 90, 25.1327, 201.0619   3. 180, 28.2743, 113.0973

11.6 – Spherical Tessellations

1. 36/720, 0.05, 1/20   2. 90/720, 0.125, 1/8   3. 180/720, 0.25, 1/4

4. 120/720, 0.1667, 1/6   5. 60/720, 0.0833, 1/12

11.7 – Spherical Triangle Congruency

1. D   2. B   3. C   4. A

11.8 – Spherical Triangle Measurement

1. 4.7124   2. 11.7810   3. 14.1372

Chapter 12 – Geometric Constructions

12.3 – Geometry Tools

1. C   2. B   3. D   4. A

12.4 – Types of Construction

1. C   2. A   3. E   4. D   5. B

Chapter 13 – Geometric Proofs

13.1 – Introduction

1. B   2. D   3. E   4. A   5. G   6. F   7. C

13.3 – Terms

1. F   2. C   3. E   4. A   5. G   6. B   7. D

Appendix C – Assessment Solutions

1. C 2. B 3. A 4. B 5. C 6. A 7. B 8. C 9. A 10. C 11. B 12. A

13. B 14. C 15. C 16. A 17. A 18. B 19. C 20. C 21. A 22. A 23. C 24. B

25. B 26. B 27. C 28. A 29. A 30. C 31. B 32. C 33. A

34. B 35. C 36. A 37. C 38. A 39. B 40. C 41. A 42. B 43. A 44. C 45. B

46. B 47. A 48. B 49. B 50. C 51. C 52. B 53. A 54. C 55. C 56. A 57. A

58. B 59. A 60. C

61. C 62. B 63. A 64. C 65. C 66. B 67. B 68. C 69. A 70. B 71. A 72. A

73. B 74. C 75. A 76. B 77. A 78. B 79. C 80. A 81. C 82. A 83. B 84. C

Chapter 8 – Polyhedron Solids – Part 2

85. C   86. B   87. A   88. B   89. C   90. A   91. B   92. C   93. C   94. A   95. A   96. B

Chapter 9 – Two Dimensional Non-polytopes

97. A   98. C   99. B   100. B   101. C   102. A   103. C   104. B   105. A   106. A   107. C   108. B

Chapter 10 – Three Dimensional Non-polytopes

109. A   110. C   111. C   112. B   113. A   114. B   115. C   116. B   117. A   118. C   119. B   120. A

Chapter 11 – Spherical Geometry

121. C   122. B   123. B   124. A   125. C   126. A   127. B   128. A   129. C   130. B   131. C   132. A

Chapter 12 – Geometric Constructions

133. C   134. A   135. B   136. C   137. B   138. A   139. A   140. B   141. B   142. A   143. C   144. A

Chapter 13 – Geometric Proofs

145. B   146. C   147. A   148. B   149. C   150. A   151. C   152. A   153. B

154-156. Each of these three questions have several possible solutions. Graded by instructor or student.

Index

Color Key: Book 1   Book 2   Book 3   Book 4

Term		Page(s)
Acute angle		33-34
Acute triangle		38
Adjacent angles		35
Adjacent side		16
Algebraic Postulates		40
Alternate angles		36
Alternate exterior angles		38
Alternate interior angles		38
Altitude		21, 42, 17, 60
Angle		4, 25, 2
Angle bisector		35, 21, 42
Angle measurement		29
Angle rotation		25
Angle side		25
Angle trisector		43
Annulus		40
Annulus sector		41
Antidipryramid		21
Antipodal points		55, 2
Antiprism		21, 25
Apeirohedron		7
Apeirotope		7
Apex		16
Apothem		19
Approximately equal		27, 38
Archimedean solid		65-67, 1, 4-6
Archimedes		4
Area		8, 19, 2
Area of a circle		36
Area of a sphere		55
Assessment		50-69
Augmented		8
Auxiliary lines		39
Axiom		18, 39
Ball		55
Base		21
Bi- or Di-		8

Hexagonal trapezohedron	21
Hexahedron	60, 61
Hexateron	5
Hexecontrahedron	60
Hexeract	5
Hilbert, David	20
Hollow shape polyhedron	59
Horizontal axis	26
Hyperbola	35, 36, 45, 46
Hyperbola asymptote lines	45, 46
Hyperbola axis	45, 46
Hyperbola center	45, 46
Hyperbola conjugate axis	45, 46
Hyperbola focus	45, 46
Hyperbola focus distance	46
Hyperbola transverse axis	45, 46
Hyperbola vertex	45, 46
Hypercube	2-4
Hyperplane	1
Hypotenuse	36
Icosahedron	60, 66, 67, 1, 2
Icosidodecahedron	60, 66, 4
Icositetrahedron	60
Incenter	34, 42
Incidence axioms	20
Incircle	19, 34, 42
Included angle	25
Indirect proof	39, 42
Initial side	25
Inradius	19, 2
Inscribed Triangle	21
Inscriptible quadrilateral	51-53
Interior angle	16
Interior angles	37
Interior of a circle	36
Irregular polygon	16
Irregular prism	24
Irregular pyramid	17
Isogonal polyhedron	65
Isohedral polyhedron	65
Isometry	28
Isosceles trapezoid	8, 50-53
Isosceles triangle	35, 36
Isotoxal polyhedron	65
Johnson solid	65, 67, 1, 8-12